Introduction

The Science Museum London, at the heart of the National Museum of Science and Industry, is a unique institution. Its collections record an event of outstanding importance in human history, the emergence of the first industrial society made possible by the blossoming of science and technology. No other museum in its field offers collections so rich or diverse nor such a wealth of material evidence fundamental to an understanding of the modern world.

Those of us whose privilege it is to look after these collections try to share with our many visitors the insight and enjoyment that comes from daily contact with three-dimensional history. This book is part of that process. It is a small celebration of the Science Museum for everyone who has enjoyed a visit here and would like to take a little of that experience away with them. I hope it will recreate something of the excitement – and understanding of the achievements of our forebears – which we feel as custodians of so much first hand evidence of how the world came to be the way it is.

Neil Cossons

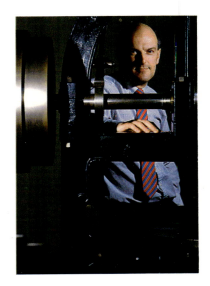

Neil Cossons

Director,
The National Museum
of Science and Industry

About this book

This book is about the Science Museum and its remarkable collections. Spanning many centuries and cultures, these collections record in material form an extraordinary human progression – from a world based on the simplest tools and techniques to a technologically advanced society, where life for many people is in many ways far richer than in the past, yet wholly dependent on the fruits of modern industry and science.

Drawing on the evidence in the collections, the book's four chapters explore interlocking themes:

The oldest surviving tinned food can, 1823.

Industry page 3
An age-old human activity, industry has increasingly become the agent which makes available to many people the products which science makes possible.

Science page 13
In a few centuries science has not only changed the way we think about the world. Through its interplay with industry it has changed the world itself.

Society page 27
Society is the seed-bed in which science and industry may grow and flourish, and is itself continuously reshaped by developments in science and industry.

Museum page 39
By maintaining and displaying its collections, the Science Museum promotes an understanding of the history of science and industry and their effects on our lives.

Prism used by
Sir William Herschel
in the discovery of
infra-red rays, c1800.

Scientific glassware
from the 10th-12th
century.

Nameplate of *Rocket*
locomotive, 1829.

Advanced industrial
product of the late
twentieth century – an
RB211 aero engine by

Industry has been an expanding human activity for thousands of years. Its practitioners may be individual craftworkers or employees of multinational companies, but they have always possessed assets not shared by the rest of society. Two of those assets, perhaps the most important, are richly represented in the Science Museum: special equipment and the know-how to use it.

Original industrial equipment – tools, machines and the products they made – is preserved in the museum itself and in its stores and out-stations. The know-how is in the learning of the curators, and in the books and memoirs, pictures and company archives, which are in the Science Museum Library.

Tools and machines

Preserved, where possible in working order, production equipment can reveal the state-of-the-art in many industries at many stages in their development. By studying it, and its products, we

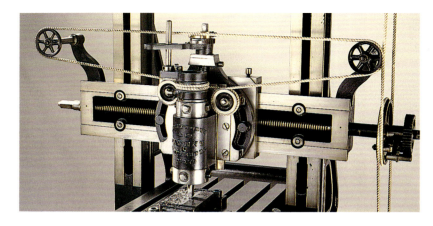

I

2

Joseph Whitworth's 1842 machine for planing metal (1) impresses with its rugged self-confidence. It helped reduce one hundredfold the labour cost in making a flat metal surface – the basis of many machines. Whitworth (2), a fanatic for precision, was one of the engineers who earned for Britain the accolade 'Workshop of the World.'

To advertise his art, an unknown cabinet maker of the eighteenth century lovingly decorated this toolchest with inlaid woods. The tools were simple, little more than specialised extensions to his hands. To use them well, much skill and a long apprenticeship were required.

can estimate output and efficiency, and assess the degree of strength and skill demanded of the people who used it.

For many generations simple tools, used with skill, would often suffice. By the eighteenth century, hand tools were being replaced by robust machines which brought greater precision but did not require greater skill from the operator. The precision was in the tool itself.

With high precision there came a new possibility: components so accurately machined that if one was defective it could easily be replaced by another made on the same machine. From there it was a small step to mass production and a society where hand-made things have become a rarity.

Every tool and machine in the museum's collections was itself made using other tools and machines. Tools, and the manufacturing machines which are their descendants, are a key to the modern world. Without them mankind would still be in the Stone Age.

1

2

Part of the world's first large-scale mass production enterprise, this shaping engine (1) was installed at Portsmouth Dockyard in 1804. Half-formed wooden pulley shells can be seen inside the machine, one of a set which could turn out 100,000 pulley-blocks a year (2,3) supplying Nelson's fleet and others for more than a century afterwards. This pioneering enterprise replaced 110 skilled men with ten unskilled ones, providing a foretaste of the economic and social pressures that widespread mass production would later unleash. The all-metal machines were devised by Marc Brunel (father of the more famous Isambard) whose autograph (4) comes from a letter in the museum's archive collection.

3

A tool to make tools, this automatic machine of the 1980s drills fifty holes in five components in two minutes. The operator's task is simple: load the work and press the button. The machine itself does the rest, using all the human skill that went into devising and setting it up.

4

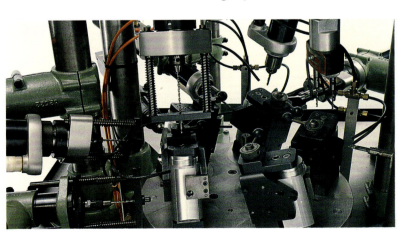

Machines and power

'I sell ... what all the world desires to have – POWER,' said Matthew Boulton, James Watt's partner in the steam engine business. After nearly two thousand years of the water-wheel, steam power began to turn the wheels of industry at the end of the eighteenth century, acting as a go-between to unlock the energy in coal and set it to work. Twentieth century factories are still steam-driven, but today the steam is confined to a power station and its energy is delivered by electricity.

The many energy-converters in the Science Museum's collections reaffirm an extraordinary engineering success story, of increasing power and efficiency combined with greater reliability, in engines which have grown lighter, smaller and easier to use. By 1803 the

1

2

Work-horse of a **Tyneside colliery** (1) for half a century, *Puffing Billy* (2) is probably the world's oldest surviving locomotive. It was built in 1813, after its constructor, William Hedley, had convinced himself by full-scale tests that a locomotive could grip the rails hard enough to haul a heavy train.

1

A gentle giant, the 1788 steam engine by **Boulton and Watt** (1) was a factory power unit capable of driving 43 separate machines. It incorporates many of the improvements in steam technology for which **Watt** is justly renowned. **Matthew Boulton** (2) provided the commercial backing and **James Watt** (3) the mechanical inventiveness in a partnership whose products soon became a byword for workmanship and innovatory design.

2

3

steam engine was compact enough to be put on wheels, and to drive those wheels itself. A century later the internal combustion engine combined high power with low weight to a degree which made the aeroplane a practical possibility.

The liberation of mankind from toil, begun by the steam engine, continues today. In technically advanced countries motors of various sorts have largely done away with muscle power, driving all the other machines and gadgets in homes and factories. Machines which handle energy, together with the industries that provide their fuel, have become crucially important to society. Without them, everything else grinds to a halt.

Machine tools and steam power were two of the driving influences in that period of accelerating change

▶ page 9

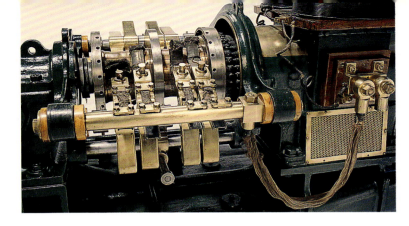

Electric generator by Sir Charles Parsons & Company: built in 1901, it helped initiate a century in which electric power would transform industry and society around the world. Steam turbo-generators still produce almost all our electricity.

Despite its business-like appearance, the 1918 Bentley rotary engine is a technological dinosaur. Built to power fighter-planes in World War I, it packed a lot of muscle into a small space, but was soon displaced by engines of less unusual design.

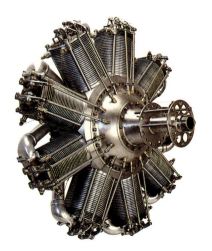

Worm's-eye view of a satellite launcher. This eight-engined *Black Arrow* rocket, two metres in diameter, was to have put Britain's second satellite into orbit in 1971, until changing government policy grounded it. Rockets outclass all other motors in their brash display of raw power.

1

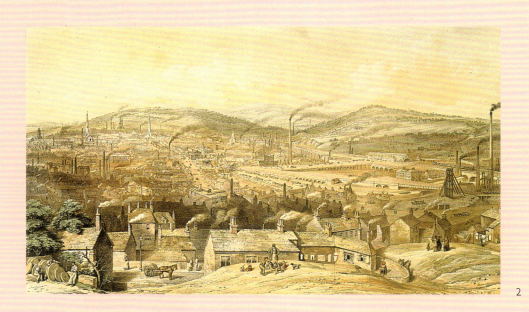

2

3

Changing perceptions of industry and its effects are reflected in three images from the museum's picture collection. In *Coalbrookdale by night* by de Loutherbourg (1) the glow of many furnaces rebounds in smoke, adding drama and menace to what is still, in 1801, a rural scene. Half a century later, industry is based more on steam and less on water power, giving rise to the urban way of life. Sheffield, in William Ibbitt's panoramic view (2), boasts mills with fifty smoking chimneys, a railway, gas lighting and impressive civic buildings. Lowry's nameless manufacturing town of 1922 (3) also has chimneys, mills and a gas lamp, but delivers a quite different message.

which has come to be known as the Industrial Revolution. It began in Britain in the eighteenth century, and transformed a largely agricultural country into the world's first industrial nation in little more than a century.

A turning-point in history

The South Kensington Museum, from which the Science Museum developed, was founded in 1857 when that transformation was at its height. It sprang from the confidence and prosperity of a generation for whom Britain's manufacturing industry led the world.

The museum is both a child of the Industrial Revolution and a record of it. As the national museum of industry in the country which 'invented' industrialisation, it is uniquely endowed with first-hand evidence of what happened at one of history's major turning-points.

Much of that evidence concerns the discovery of new processes and the invention of new machines. Such

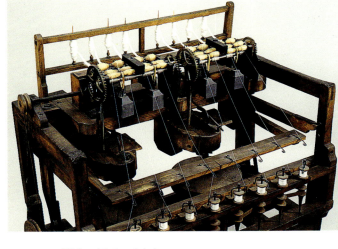

Richard Arkwright's spinning frame (1) helped to take the spinning process out of the home and into the factory. Pairs of rollers replaced nimble fingers for drawing out thread, while horse or waterwheel supplied the driving power. **Of Arkwright himself (2), James Watt said, 'he is … the most conceited, self-sufficient man.'**

2

1

1792

When official coinage became scarce in the 1780s, industrialists began to mint their own. Ironmaster John Wilkinson was bold enough to replace the king's effigy with his own (1). The second token, dated 1792, commemorates the recently opened Iron Bridge, which today still spans the River Severn at Coalbrookdale (2).

2

technological innovations were important among the many ingredients which fuelled the Industrial Revolution.

Some inventions were the result of dogged persistence, others came from flashes of brilliant intuition or by following up a piece of good fortune. But while it is often the inventors who leave relics for museums to collect, their ingenuity alone has seldom led directly to successful innovation. The enterprise of others has been equally important – industrial backers, managers, publicists and financial risk-takers – drawing of course on the labour of the work-force. Such people leave less collectable evidence of their activity, but ensure that the inventor's original idea can take root and mature into an influential process or product.

Science and industry

The Industrial Revolution is still going on, propelled by innovations that have sprung from the scientist's

A scientist's response to a spate of explosions in coal-mines, Sir Humphry Davy's safety lamp (1) was first tested underground in 1815. Engineer George Stephenson proposed a lamp of very different design (2). The Marsaut lamp of the 1880s (3) proved one of the safest and saw many years' successful service.

James Nasmyth's 1871 painting of a steam hammer celebrates industrial production on a heroic scale. In a scene resembling grand opera, slave-like humans labour to satisfy the machine's demands, man-handling a massive iron bar between the hammer's jaws. Engineer as well as artist, Nasmyth was famous as a builder of steam hammers.

laboratory. The new science of electricity first bore fruit, for example, with the electric telegraph in the 1830s. In a century and a half it has blossomed into the complex world-wide communications industry.

Science has brought us new materials, such as aluminium, first extracted in quantity in the 1880s, and fully synthetic plastics, of which Bakelite in 1909 was the first. From the chemical industry, too, has come a growing range of products of ever-increasing complexity – fertilisers, insecticides, drugs and medicines.

Late in the twentieth century the smoke-stack image of industry is beginning to fade. Science-based

2

A sample of the first synthetic dyestuff ever made (1), inadvertently created in 1856 by an eighteen year old chemistry student, William Perkin, while attempting to produce something else. The shawl (2) was dyed in Perkin's Mauve a few years later. Other artificial dyes followed, making possible the colourful products in today's shops and homes.

The Marconi short-wave radio transmitter of the 1920s was a product of the fast-growing electronics industry. Designed by one of Marconi's own engineers, it was used to beam messages to the USA and South America from a Post Office radio station in Britain.

industries like electronics and bio-technology present a new picture of 'green field' factories which are as clean and quiet as a laboratory, and staffed by highly-trained technologists.

Yet in spite of many changes industry is still what it has always been – a service which makes the ideas and talents of a small number of people available to many. More and more, it has also become the principal way in which the discoveries of science come to affect our lives.

For the museum, new industries present a continuing challenge. Their equipment and products are less attractive to look at than the relics from earlier days, their function and meaning less easy to explain. Changes often come thick and fast as the cycle from introduction to obsolescence grows shorter. Today's significant innovation rapidly turns into tomorrow's historical evidence, and the choice of what should be preserved grows harder.

1

The production of insulin by fermentation is recorded in a painting by Alan Stones (1). It shows the Dista Products plant in Liverpool, first in the world, in 1982, to use genetically-manipulated bacteria in the production of medicines. The resin column (2) is one of several employed in separating and purifying the insulin.

2

Like all components for use in space, this miniature tape recorder demanded the highest production standards. Special clothing is worn to protect vital components from contamination by dust and grease from people. The recorder was built as a spare for a British satellite, *Ariel VI* launched in 1979.

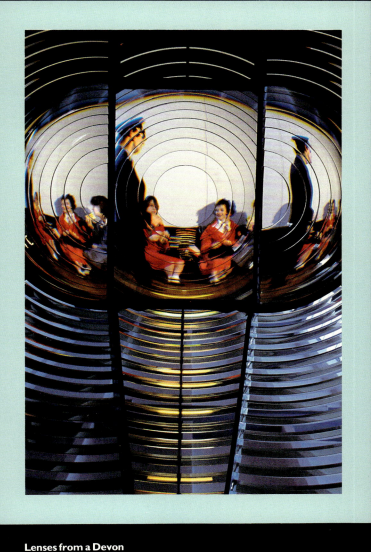

Lenses from a Devon
lighthouse, made to a
pattern devised by
French scientist
Augustin Fresnel,
produce three images
of a single scene.

In a few centuries science has become one of mankind's supreme cultural enterprises, transforming the way we think about the world. More than this, through its interplay with industry, science has reorganised society, directing the way people live their lives in the technically-advanced countries of the world. Often, through developments in medicine, public health and warfare, it is science which decides for each of us when we live and when we die.

Observation and experiment

Three centuries ago there were only a handful of paid scientists in the world. Science was mainly something done by amateurs, often wealthy

A book illustration by Nicholas Copernicus (1) is the first to show the earth in orbit round the sun. Published in Latin in 1543 and called *On the Revolution of the Heavenly Spheres,* the book began a revolution in human thought, displacing mankind from the centre of the universe. The working model (2) reflects the new viewpoint, reproducing the true movement of the earth and moon around the sun. It was made by John Rowley, *c*1712, and belonged to his patron the Earl of Orrery. Such models have been known as *orreries* ever since.

2

Christopher Cock, one of the first optical instrument makers in London, made this telescope in 1673. As their appearance suggests, many early telescopes and microscopes were simply playthings for the well-to-do, but they also symbolize a period when science began to open new worlds for exploration.

gentlemen who would meet regularly to ask questions and discuss new findings. The one essential rule of such gatherings was that the answer to their questions should be sought in nature itself. More than ever before, the way to new knowledge was to be through observation and careful experiment.

It was an approach which would make science an astonishingly successful human activity in the years to come.

Evidence of the activity of scientists, and of the way it has grown and changed through the years, is preserved in the museum's splendid and varied science collections. Instruments and equipment which investigators have left behind hold

▶ page 18

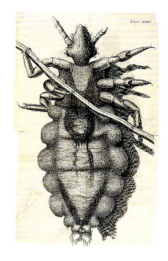

1

With a microscope we can 'peep in at the windows' of nature, wrote Robert Hooke, a contemporary of Isaac Newton. The illustration, of a louse clinging to a human hair (1), is an example of what he saw. Hooke drew it for his book *Micrographia*, published in 1665 – a classic which so fascinated the diarist Samuel Pepys that he sat up reading it until two in the morning. The microscope (2), from c 1675, is similar to the one used by Hooke and illustrated in *Micrographia*.

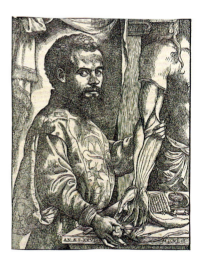

In a woodcut first published in 1543, Andreas Vesalius illustrates the working of the human arm. Vesalius was professor of anatomy and medicine at the University of Padua in Italy. His youthful confidence reflects the mood of an era in which ancient wisdom was being challenged by first hand observation.

2

A small selection from the museum's sumptuous instrument collection recalls an era when the boundaries of science and art were less sharply defined. Instruments were constructed with lavish care and artistry, often by craftsmen of whom little is known. Spanning three centuries, the examples on these pages chart an evolutionary path in which complexity and precision increased as instruments progressed from being accessories for gentlemen into tools for professional scientists and engineers.

1

3

2

The astrolabe (1) was an aid to surveying and navigation, and in the casting of horoscopes; the one shown here is Dutch, and dates from about 1570. The superb circumferentor (2), by Joannes Macarius and dated 1676, was a surveyor's instrument for measuring angles. The sector, protractor, and portable sundial (3) are part of a set of drawing instruments made by Dominicus Lusuerg in Rome in 1701. Advertised on the trade card (6) are some of the items supplied by Thomas Wright, a London instrument maker, early in the eighteenth century. Jesse Ramsden, possibly the greatest of English instrument makers, built the large theodolite (5) in the late eighteenth century; used for accurate surveying over long distances, it owes its precision to the use of telescopic sights and to Ramsden's extraordinary skill. In the nineteenth century, instrument making moved from workshop to factory. Florenz Sartorius founded his balance-making business in Germany in 1870. His short-beam balance (4) can distinguish weights which differ by as little as one part in a million.

4

5

6

clues to the ways discoveries are made, and stand as a memorial to the achievements of pioneers.

As science has developed, its nature and organisation have changed. By 1800 it was an international activity, and a hundred journals were available to spread the news of the latest findings. Scientists were increasingly to be found in universities. But important discoveries could still be made by amateurs working with their own simple equipment in their own private laboratories.

A 'useful art'

A century later, science had grown into a recognizable and respected profession, mainly done in purpose-built university laboratories. Equipment belonged to the institution, not to the scientist, and instruments were no longer the works of art they had been in earlier days. The trend can be traced

2

A well-equipped chemical laboratory of the 1890s signals the growth of science as a profession. The reconstruction includes original fittings from the Government Chemist's Laboratory in London, one of the first buildings designed specifically for chemists to work in.

Heated and sealed in the 1860s, this flask (1) has remained sterile ever since. It was used by Louis Pasteur in experiments which proved that tiny organisms in the air can cause decay. His work led to an understanding of the role of germs in causing disease, and to the introduction of antiseptics. The statuette (2) shows Pasteur holding a similar flask.

through apparatus of the period; it becomes visibly more complex, more specialised and more severely functional as the years go by.

From the start many scientists had seen their business as more than simply the pursuit of knowledge. Science was also a 'useful art', to be applied to invention and the improvement of manufacturing processes. Electricity was one science which began to bring practical benefit in the nineteenth century. Medical science was another, leading to the prevention of disease as well as to new methods of treatment. Science was coming to be seen more and more as a force to improve people's lives.

Science was carried into the twentieth century by its success in the nineteenth. The number of scientists continued to grow at an ever-increasing rate, both in

▶ *page 21*

Designed by a Professor of Natural Philosophy, this 1872 'computer' tackled a very practical problem – predicting the rise and fall of the tide in a harbour. Like many scientists before and since, its creator, Sir William Thomson (later Lord Kelvin), was happy to target his research on matters of direct practical concern.

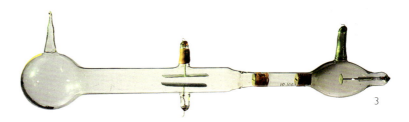

Vacuum tubes, chosen from many in the collections, show some of the varieties developed in the late nineteenth century: (1) a tube devised by Sir William Crookes to show how cathode rays could be deflected by a magnet, (2) a lightbulb converted into one of the first radio valves by Ambrose Fleming, and (3) a tube which J J Thomson used in his experiments to discover the electron. Shown in action (4) is a tube which contains low-pressure gas, made at the turn of the century purely for show. Research with vacuum tubes, made possible by improvements in vacuum pump technology, gave rise to atomic physics and to applications which include X-rays, radar and the modern television tube.

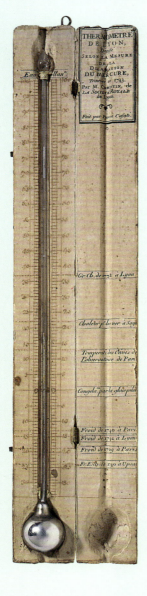

Measuring instruments have long been essential in trade and commerce, replacing estimates – and arguments – with numbers. In the development of science too, measurement has been a crucial aid to understanding. As Sir William Thomson explained in 1883: 'When you can measure what you are speaking about and express it in numbers, you know something about it.' With the selection of instruments on this page, time intervals can be measured with a set of sand-glasses dated 1720 (1), temperature with a 200-year-old thermometer, one of the first to use the Celsius scale in the way we use it today (2), and blood pressure with a portable instrument from 1890 (3). The weights (4) are from a set, created by order of Queen Elizabeth I, which formed the 'standards of the Realm' from 1588 until they were replaced in 1824.

universities and in research laboratories set up by industries and government departments. By the 1960s it could be said that, of all the scientists who had ever lived, four-fifths were alive then, and the same is true today. Science has become an industry in itself.

Equipment and models

Equipment has grown ever more elaborate, and hugely expensive. In their modern versions, the microscope and telescope are still important, but they are now just two items on a much longer list of tools for investigation and analysis. Twentieth-century additions to that list include radio-telescopes, body scanners, and particle accelerators. Some equipment is now so complex that only governments can afford it, some projects so costly that they have become multinational ventures.

Identifying, among so much that is new, the key developments which will later be seen to have major significance, is one of the science curator's most challenging tasks.

There has always been more to science, of course, than simply

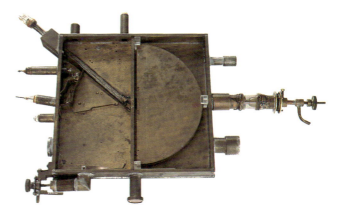

Lawrence and Livingstone's 1932 cyclotron was used to bombard and split atomic nuclei. Its D-shaped electrode, just visible here, is a modest 28 centimetres in diameter. Half a century later, descendants of this machine had grown to be the most costly tools of pure research on Earth, with diameters measured in kilometres.

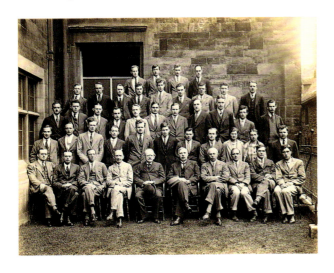

Researchers in 1928 at the Cavendish Laboratory, Cambridge – a mecca for physicists drawn by such eminent names as Sir J J Thomson and Sir Ernest Rutherford (seated, with hats). Of the years ahead J D Cockcroft (fourth row, second from right) would later say: 'We were living in the Golden Age of physics, so rapidly did discoveries come along.'

'Big Science' in 1937, the million-volt Cascade Generator provided high voltages for research equipment in Cambridge. With its distinctive silhouette, necessary for insulation, and the power to deliver imitation thunderbolts, it inspired a generation of science fiction writers and illustrators.

observing and doing experiments. Like a detective, the scientist needs insight and imagination, to know the right questions to ask, to interpret unexpected results, and to tease out patterns from a mass of data.

To help with this task, scientists construct models of many kinds – man-made versions of reality. Sometimes a model is a theory, to be tested by further experiment. Often it is a real piece of hardware, giving substance to what has been discovered about the structure of a molecule, the mechanics of

This Scanning Microanalyser is the prototype of a powerful tool for scientific detective work. Capable of revealing the chemical composition at any tiny point on the surface of a specimen, it developed from work in university and industrial laboratories in the 1950s, and went into commercial production in 1960.

The satellite *Ariel I* was a scientific space explorer which carried into orbit experiments devised by four British universities. Shown here is the full-scale engineering model, used for tests. The flight version was launched on 26 April 1962, and burned up on re-entry 14 years later.

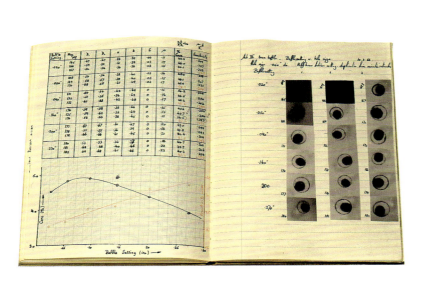

The day-to-day progress of grass-roots research is documented in a laboratory notebook from the 1960s. It records experiments done to discover how a new type of artificial fibre could best be manufactured.

the human body, or the layout of the universe.

Such models feature prominently in the museum's science collections. They are part of the essential process of science, helping researchers to refine and express half-formed ideas. Preserved for study, they give insight into the way scientists' minds work.

Models can play a second important role. Together with specially made demonstration equipment they can be used to explain the discoveries that scientists make, helping us explore for

▶ page 26

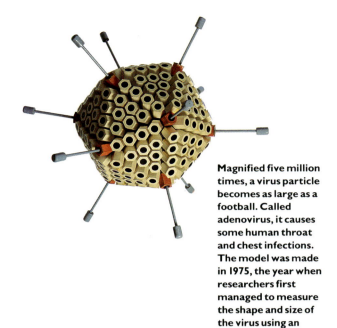

Magnified five million times, a virus particle becomes as large as a football. Called adenovirus, it causes some human throat and chest infections. The model was made in 1975, the year when researchers first managed to measure the shape and size of the virus using an electron microscope.

In the 1830s Professor Daniell, of King's College, London, used these models to show his students how atoms might group together to form the various crystal shapes found in nature. The new theory of atoms, put forward by John Dalton, had gained much ground by that time.

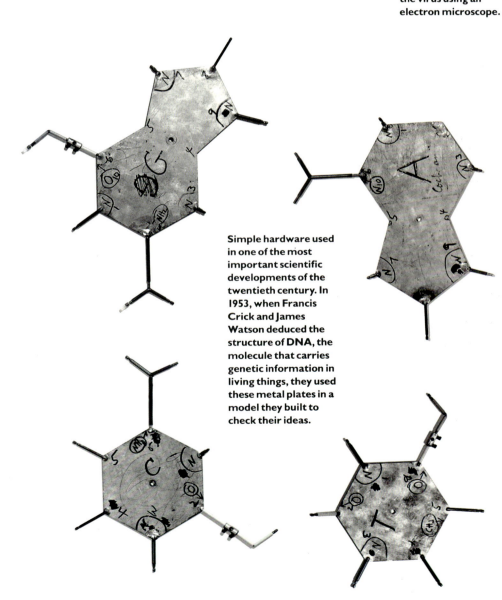

Simple hardware used in one of the most important scientific developments of the twentieth century. In 1953, when Francis Crick and James Watson deduced the structure of DNA, the molecule that carries genetic information in living things, they used these metal plates in a model they built to check their ideas.

Devices from three centuries outline the rapid evolution of calculators and computers. The common feature which makes them all 'digital' is that they work with indivisible units – beads on a wire, teeth on a cogwheel or electric pulses in a circuit.

The abacus (1) has helped shopkeepers and traders for thousands of years. Cogwheels, their movements controlled by exact numbers of teeth, began to be used in calculators in the seventeenth century. The earliest British ones date from the 1770s (2). In the nineteenth century Charles Babbage (3) stretched cogwheel technology up to – and beyond – its limits. His Difference Engine (4) was intended to calculate and print mathematical tables. Later he designed a machine which had many of the attributes of electronic computers that would not become reality for nearly another century, including the use of punched cards (5). Babbage's plans were too advanced for the technology of his time, however, and he died, dissatisfied and unhappy, in 1871.

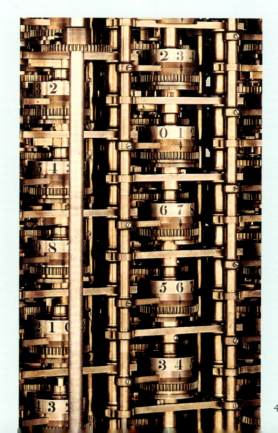

6

7

The German *Enigma*
machine (6), not itself a
computer, could turn
military messages into
code in more different
ways than there are
atoms in the universe.
In Britain, experience
in building electronic
code-breaking
machines led to the
design, in 1945, of *Pilot
Ace* (7). With eight
hundred radio valves,
it was one of the first
real computers,
operational in 1950.

Valves were
soon replaced by
transistors,
miniaturised into the
components which
make 1980s computer
technology so
distinctive (8). By then
computers were
performing many
more tasks than the
simple 'number-
crunching' feats they
began with.

8

ourselves the new perspectives which science has created.

Two centuries ago King George III commissioned instruments and equipment which could be used for scientific demonstrations, and which are now held by the museum. They form a uniquely significant focus for all the science collections, for their function can be seen as one which remains to this day a central purpose of the Science Museum – to collect and display things which will help its many visitors understand what science is about, what scientists do, and how their discoveries have changed the world.

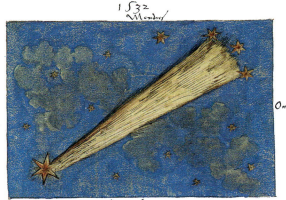

In the sixteenth century, when this small watercolour was painted, a comet was a much feared omen. Since that time, for many people and in many areas of experience, the rise of a scientific world-view has replaced terror and superstition with wonder and understanding.

Demonstration equipment constructed for King George III consists of a bell in a jar from which air can be removed with a pump. It allowed the King to share his enthusiasm for 'natural philosophy', in this case by showing his family and friends that removing the air makes the bell sound quieter.

1

The *Apollo 10* spacecraft (1) carried astronauts around the Moon and back in May 1969, a rehearsal for the first lunar landing a few weeks later. Made possible by developments in rocketry and electronics, the Apollo programme reflected, and reinforced, a confidence in the power of technology that was typical of the time. Through their spacecraft window the *Apollo 10* astronauts saw and photographed their distant home (2), helping humanity acquire a new perception of 'Spaceship Earth' which is perhaps the most significant legacy of the Apollo adventure.

2

**Tramcar No.585 helped
millions go about their
business in the city of
Glasgow in the first
sixty years of the
twentieth century.**

The Science Museum's collections fulfil two roles. In addition to shedding light on the way science and industry have developed, they bear witness to the effects which science and industry have had on the structure of society and on people's lives.

This social influence is reflected most directly, not in things that are rare and special, but in items which the museum has acquired because they are typical, representative of what people have owned and used – the technology of everyday life. **Displayed for all to see, the changing technology of home and work, travel and recreation can help each of us to build a personal picture of what the past was like. Such pictures**

2

Made and used more than a century apart, these instruments represent great strides in preventive medicine. The nineteenth century vaccination kit (1) **contains small sharp blades used to pierce the patient's skin so that smallpox vaccine could be applied. By 1980, a gas-driven gun** (2) **could place vaccine through the skin directly. Using methods like these, the great scourge of smallpox has been vanquished and the grip of other diseases weakened.**

Good drainage and clean water supply are crucial to health. The will to provide them and the means to do so emerged in Britain in the nineteenth century. Whilst potteries mass-produced standard ceramic piping (this is a travelling salesman's sample), engineers like Joseph Bazalgette in London undertook huge drainage schemes.

X-rays were a landmark in medicine. Discovered by Röntgen in 1895, they were greeted with public wonder, and quickly adopted by doctors as a vital new tool for diagnosis. When this X-ray was taken, early in 1896, it was the first in Britain to show part of the human body.

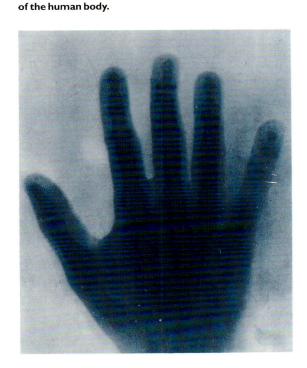

contribute to a knowledge of history without which, as historian Arthur Marwick has expressed it, 'man and society would run adrift, rudderless craft on the uncharted sea of time'.

Health and life

Of all the changes that industrialisation and the rise of science have brought, none is more directly relevant to everyone than the improvement in health and life expectancy. Better understanding of the way disease is spread has contributed to this, as have new techniques for preventing and treating illness, for distributing clean water, and for the production of food. Above all it is the higher living standards which only an industrialised society seems able to support that have made the most difference. Though industrialisation

From the 1960s, the Pill gave many women the chance to use effective contraception for the first time. The fear of unwanted pregnancy was removed. Although no longer viewed as a panacea (the side effects of long-term use are not fully understood) its introduction prompted profound social and psychological change.

The agony of toothache and primitive treatment alike have receded as science has come to the dentist's aid. By the 1930s, the use of anaesthetics was well established – here the anaesthetist administers gas. The chairside dental unit brings drill, implements, water and light within easy reach. Sterilisation and X-ray equipment are also to hand.

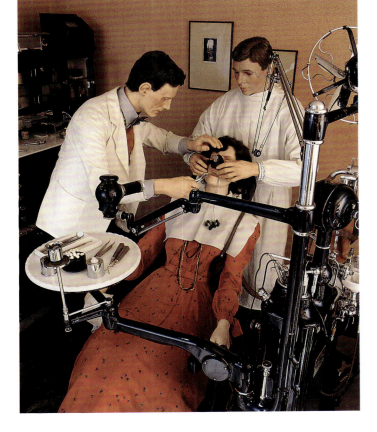

Steamships conquered wind and tide, and Brunel's *Great Eastern*, 1858, was the shape of things to come. She first crossed the Atlantic in 1860; many other steamships followed, laden with immigrants to populate new lands. This great ship ended her days laying submarine cable, promoting other links across the sea.

Railways transformed travel, industry and settlement alike. Their growth boosted the industrial revolution to a new, more active, phase. Britain's thriving Victorian economy was based on their might. *Caerphilly Castle*, retired in 1960 after nearly three million kilometres of service, embodies the spirit of the railway age.

brings far-from-trivial ills of its own, among its undoubted benefits is the fact that, in Britain at least, twenty times fewer children now die before the age of five than was the case a century ago.

A more mobile society

With higher living standards have come greater opportunities for travel, made possible by the harnessing of new power sources in self-propelled vehicles.

Passenger railways came first, in the 1830s. Before then any journey of more than a few miles had been a rare adventure. Soon rail travel would be within the means of many.

By the 1890s developments in electric motors made possible the sort of rapid transport systems which allow people to live in one place and work in another – leading to cities with suburbs and to the commuter way of life.

The internal combustion engine, compact and running on a convenient liquid fuel, drove the first motor vehicles in the 1880s. Few people enjoyed the benefits, however, until after 1910 when motor-buses became common. Only after World War 2 did the

Cycling caught on in a big way after the launch of the *Rover Safety Bicycle*. Its classic design, seen to good advantage in this prototype of 1888, with the added comfort of pneumatic tyres (introduced by Dunlop in the same year), proved a recipe for success. Bicycle manufacture boomed, providing welcome personal transport here and overseas.

private motor-car become a feature of many British households. Soon cities, helped to grow and flourish by the earlier boom in public transport, would be increasingly threatened by the demands of private transport.

Most recent of the transport revolutions has been the arrival of affordable air travel, the result of improvements in jet engines and the advent of 'jumbo' aircraft in the 1970s. With its world-wide network and computer-age communications it has turned tourism into a major industry, and made air travel a regular event in the lives of millions.

The evolution of today's highly mobile society is mirrored in the museum's comprehensive transport collections. Few developments in technology have offered so much, in terms of individual freedom, and few have demanded so much in return, in environmental impact and damage to human lives.

While transport was reorganising people's lives outside the home, equally far-reaching changes were going on indoors. If we removed from a modern home all the things

▶ page 34

LIFE WORTH LIVING.
With an "Austin Seven" you can spend it where you will, for travel costs you about 1d. per mile with your jolly young family. The "Austin Seven" makes happiness inevitable, and it is a smart little turnout to be proud of.

The *Austin 7*, launched in 1922, was Britain's first mass-produced small car. Before then, vehicles were craft built, and only for the rich. The *Austin 7*, shown here in the manufacturer's catalogue, was sold under the slogan 'The Motor for the Million', and helped bring car ownership to a much wider public.

Picturesque it may be, but the 1958 Ford *Edsel* was a commercial flop. Giving only fifteen miles to the gallon, the *Edsel* guzzled precious fuel. Modern vehicles are designed to make better use of scarce and expensive resources.

The *Trident 3B*, seen here as it joined the museum's Air Transport Collection at Wroughton airfield in Wiltshire in 1986, was one of the post-war jets that brought foreign travel within most Britons' reach. The massive growth in air travel since the 1960s has helped bring global awareness to us all.

An assortment of domestic and industrial items demonstrates a range of different materials doing jobs for which their physical properties make them particularly suitable. It underlines the important part which an understanding of materials has played in the development of the modern world.

The ability to work existing materials – and to create new ones – has long been taken as an indicator of technical progress. First came the stone age, then bronze and iron. Continuing the sequence, archaeologists of the future might well christen our own time the 'plastics age'.

Stone is represented by a crude oil lamp from the Shetlands (1). The gas burner (2) is of porcelain, enabling it to

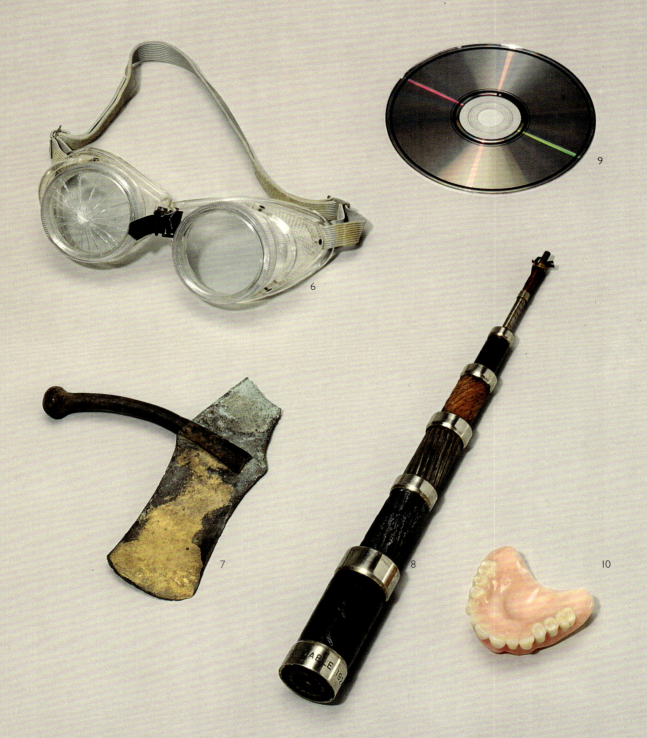

be used in places where metal would corrode. Ivory was used for the 1649 portable sundial (3), and wood for the nocturnal, an instrument for telling the time by the stars (4). The engraved printing block (5) also takes advantage of the properties of wood. The goggles (6) are of safety glass, and show that it does not splinter, even when cracked.

The way in which metals can keep a cutting edge is illustrated by a bronze razor from the 14th century BC (7), while their electrical and magnetic properties are indicated by a sample which shows the structure of the 1924 telegraph cable under the Atlantic (8). The compact disc (9) and denture (10) are two of the very many modern products for which plastics are the most appropriate material.

that would not have been there a century ago, only the most basic necessities would remain.

Domestic technology

In 1890 heating and cooking were still almost entirely fuelled by wood or coal; lighting was by candle or gas. Electricity in the home was still a novelty, and would not become commonplace for nearly half a century. The small electric motor had not been developed, so labour-saving devices were unheard of – no electric washing machine, vacuum cleaner, or fridge.

In the 1890s home there were none of today's plastics and artificial fibres; carpets were an expensive luxury. Lacking aluminium and stainless steel, kitchen equipment was crude and unwieldy by the standards of today.

In the family medicine chest a century ago there would be no

A sewing machine was probably the second complex mechanical device, after a clock, which most homes acquired. Whilst it banished much onerous hand-sewing at home, the sewing machine also led to increased use of sweated labour in the tailoring and dressmaking trades. This design was manufactured by William Jones of Manchester for thirty years from 1879.

As the last century closed, gas heating was beginning to banish messy coal fires with all the work they demanded in preparation and cleaning. This early gas fire from the mid-1880s sported tufts of asbestos which glowed in place of the fireclay radiants that were used later.

Symbol of the ease which electricity has brought to everyday life, the switch is everywhere, but is rarely as elegant as this turn-of-the-century delight. Electrical installation took time: even by 1930 fewer than half the homes in Britain were wired up. The major spread was between then and the 1950s.

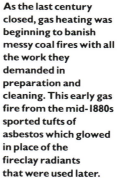

aspirin. Diseases like tuberculosis, scarlet fever and diphtheria would be feared for many years to come.

Domestic technology, much of it from the past hundred years, forms a significant ingredient in the collections. To see familiar things as they were in the past, and think about life without them, sets in perspective the possibilities – and the problems – which science-based technology brings to our own lives.

By continuing to acquire things which are special only because they are ordinary, today's museum curators will help future generations understand what it was like to live in the twentieth century. They are salting away some of the evidence that scholars will one day need to unravel the ways in which future society is shaped by present technology.

▶ page 37

When *Pam*, the first all-transistor radio made in Britain, appeared in 1956, there were already portable valve radios – though their battery demands made them expensive to run. Transistors led, before long, to miniature radios at affordable price. Sound could now go anywhere.

Thanks to satellite television, the world can reach your living room. In 1987, when this picture was taken, eavesdropping on satellite broadcasts by means of a private rooftop 'dish' was a luxury few could afford. New satellites and services were expected to reduce the cost, allowing more people to opt for the choice of viewing which satellites can offer.

Portable cameras, like this folding one from about 1914 (1), turned photography into a pastime which amateurs could enjoy. Gone was the need for bulky equipment and professional training; all could record their own history as they pleased. The charming view of two women enjoying river sights (2) was a typical result.

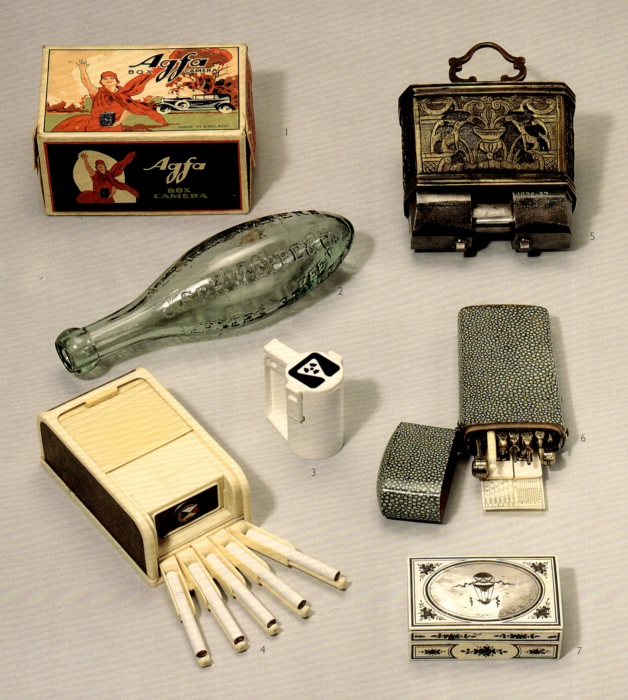

A miscellany of containers shows varied solutions to problems of packaging. The needs of what is to go inside combine with demands of function, style, cost, and the available technology to determine the form of the container.

A cardboard box from the 1930s (1) advertises the product it contains, as does the 1886 mineral water bottle (2). The modern canister, of plastic lined with lead (3), holds radioactive material for medical use, and advertises its contents for a different purpose. Also of plastic, the roll-top dispenser (4) delivered cigarettes to passengers on an ocean liner in the 1950s.

A hefty lodestone (5), from 18th century Russia, is mounted in decorative brasswork which protects it and keeps in place the iron pole-pieces which make it a more effective magnet. The case for a set of fine drawing instruments (6) is covered in hard-wearing fish-skin. The ivory trinket box (7) is French, from the 1780s. Its decoration celebrates an amazing new technology of the day – the balloon.

A century ago, many communities were still islands, cut off from the outside world. Only a few well-to-do people wrote letters or could afford to send an occasional telegram.

The wonder of the age

A hundred years of spectacular development in communications have changed all that. The telephone, radio and television, each in its day 'the wonder of the age', would still seem miraculous if we were not so accustomed to them. The telephone has allowed families to spread around the country and the globe, yet keep in touch. Newspapers and the electronic media bring into our homes impressions of the outside world which can awaken concern and help to break down prejudice.

By the 1980s images of particular relevance to the story the Science Museum tells were becoming painfully familiar. They showed how unevenly spread around the world

Sholes and Glidden's 1875 machine may seem quaint, but this is deceptive. The two inventors gave the word *typewriter* to our language, and developed the QWERTYUIOP keyboard familiar to all today. Typewriters revolutionised office work, opening up job opportunities for women which were considered an advance at the time.

The first really effective means of long-distance communication, this electric telegraph by Cooke and Wheatstone proved its worth spectacularly in 1845 by carrying the message that trapped a murderer. Telegraph lines ran alongside railways; from 1852 they also transmitted daily time checks, spreading standard Greenwich time across the nation.

In 1898 this National Telephone Company directory listed almost every subscriber in Great Britain and Ireland in a single volume. The yearly telephone rental, at £10 far too expensive for most people, included unlimited local calls – hard on users in Ruabon, with only five other names to choose from.

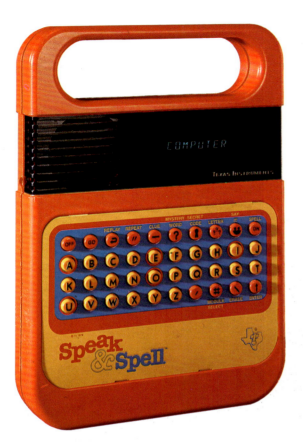

are the effects of technological development. Stages of industrialisation which are history in the Science Museum in Britain are closer to the present state of progress in some other parts of the planet. For some of the world's people Western-style industrialisation is a ladder whose lowest rung they have not yet reached.

Technology itself, in the form of global television, has made us aware of one of mankind's most formidable challenges: to find a means by which we can share more equitably amongst all nations the benefits which advanced technology can bring, realizing to the advantage of everyone the possibilities offered by the accelerating advance of science and industry which is so vividly documented in the Science Museum's collections.

In 1976, the *Cristalonic* was the first quartz watch to be powered by the sun. With a solar cell to recharge its batteries, it uses the most renewable energy of all, and points in its own small way to larger applications which may one day help reduce our dependence on fossil fuels and nuclear power.

As each new technology is introduced, new uses appear that had not been thought of before. 1970s microelectronics made possible this toy to help youngsters learn to spell. Other applications await the imagination of inventors to come.

The *Spectrum* was one of many home computers available in the mid-1980s. Information technology was transforming people's working and business lives, but, despite millions sold, it was not yet clear how great an effect the computer would have on life at home.

Items from the Plastics
Collection, chosen to
show how plastics are
used in building, catch
the eye of a visiting
student.

Combining audacious construction methods with revolutionary design, Joseph Paxton's *Crystal Palace* (1) caught the imagination of Victorian Britain. In 1851 it housed the *Great Exhibition of the Works of Industry of All Nations,* whose 100,000 exhibits included the latest products of the world's engineering industries (2) amongst which Britain was pre-eminent at the time. Six million people flocked to the exhibition in less than six months, leaving the organisers with a handsome profit.

Prince Albert (3), Queen Victoria's consort, had been a leader in promoting the Great Exhibition. His vision and influence were equally important in what was to come, for the profits of the exhibition were put towards the purchase of a nearby estate, to be dedicated for 'the application of Science and Art to industrial pursuits'.

By 1857 the first museum was established, firmly stating its role by devoting its central hall to the 'Educational Collections' (4). Since that time the South Kensington estate has blossomed into a unique educational and recreational complex – affectionately known as 'Albertopolis' – whose assets include many museums and colleges devoted to the arts and sciences.

1

2

3

4

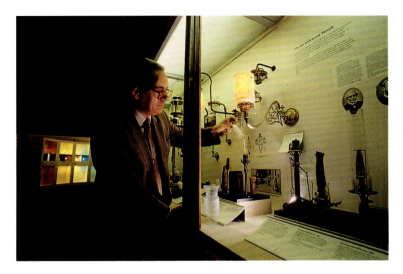

A senior curator reviews objects in his care – part of the **Lighting Collection** – while preparing to demonstrate how some of them were used. The museum assigns every new acquisition to one of more than a hundred specific collections. Collections are grouped together and looked after by specialist curators.

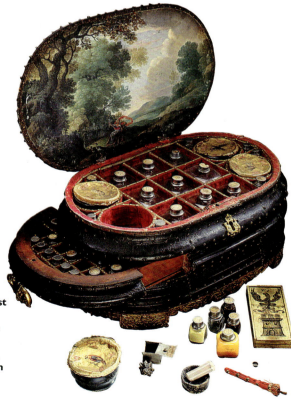

A fine sixteenth century medicine chest is just one of 130,000 objects transferred to the Science Museum from the collection of Sir Henry Wellcome in the 1970s. Part of this vast and varied collection is displayed, with more recent acquisitions, in *The Wellcome Museum of the History of Medicine* within the Science Museum.

Within the society which industry and science have helped create, the Science Museum has an important function. It promotes an understanding of the history – and present practice – of science, technology, industry and medicine, and helps its many users become well-informed citizens in a world where such understanding has never been more necessary.

To achieve its aim the Museum builds, maintains and exhibits the national collections, made up today of more than 200,000 individual items – referred to as 'objects' by museum people. To earn a place in these collections each new object must make a specific contribution.

Building the collections

One object may illustrate a significant step forward in the development of science or technology. Another may be the best example to demonstrate

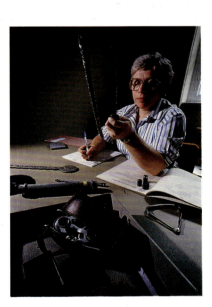

Details of a specimen of titanium, acquired to illustrate a possible new production technique, are recorded by the curator of the Metallurgy Collection. As for all new acquisitions, the information will help define this object on the museum's inventory – the database of all its holdings.

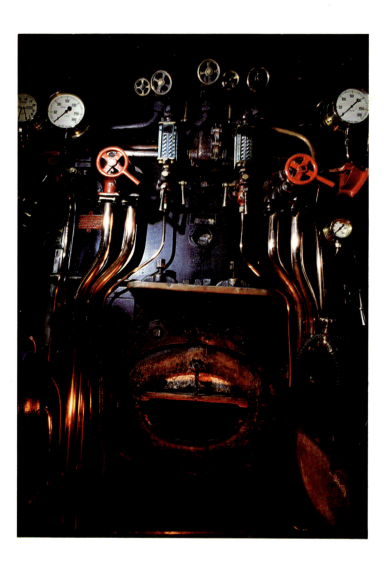

The *National Railway Museum* in York has major collections illustrating the history and development of British railway engineering, including their social and economic aspects. This is the cab of the locomotive *Mallard*, holder since 1938 of the world speed record for steam traction – 126 mph. *Mallard* was restored to working order in 1986.

The *National Museum of Photography, Film and Television* in Bradford explores the history of these three media and demonstrates how each uses technology to make pictures. Here, young visitors ride a 'magic carpet' to different parts of the world by means of the Chromakey exhibit in the *Television behind the Screen* display.

what was typically used at a particular time. A third may have important associations – often with a notable person or event.

Some objects are selected specifically to meet the needs of exhibition. They may evoke associations in the minds of visitors, be useful for explaining and illustrating a theme or principle, or simply be specially interesting or attractive to look at.

Just a few of the exhibits still on display came from the Great Exhibition of 1851. They formed part of the South Kensington Museum which opened in 1857. By 1909 its science collections were so extensive that the Science Museum became separate from the art collections, which formed the Victoria & Albert Museum. By 1928 the first part of the present Science Museum building was open.

Today the Science Museum in Kensington is headquarters – and

The *Science Museum Library* at South Kensington is not just a resource for museum staff. As the leading national institution in its field, its facilities are available to all who enquire into the history of science and technology. Here one of the librarians refers to catalogues in response to a reader's enquiry.

Science Museum Wroughton, an airfield near Swindon, is home for many of the museum's most massive objects. Here a 1934 heavyweight from the Road Transport Collection is prepared by its curator for a run at the Gala Day – one of several occasions each summer when *Science Museum Wroughton* opens its gates to the public.

'flagship' – of the National Museum of Science and Industry, an enterprise which also includes the Science Museum Library, the Wellcome Museum of the History of Medicine, the National Railway Museum in York and the National Museum of Photography, Film and Television in Bradford. In addition there are two warehouse stores in the London area and, in the west of England, the Concorde Exhibition at Yeovilton and Science Museum Wroughton, an airfield where large objects such as buses, farm machines and airliners are kept.

Research and display

Collecting and looking after objects is only a part of the work of the museum. Research and display are also essential if it is to fulfil its function. As messengers from the past, all

▶ *page 46*

The Science Museum's *Concorde* has an exhibition to itself, alongside the *Fleet Air Arm Museum* at Yeovilton in Somerset. *Concorde 002* revealed its famous 'droop nose' to the world in 1969. After 439 flights it finally touched down at Yeovilton in 1976.

Reserve collections – objects not currently on display – are looked after in several large stores, such as this one at Hayes in Middlesex. They are available to those with a specialist interest, and may form part of future exhibitions.

Day by day running of the museum involves more than four hundred staff in an unusually broad range of duties, caring for the collections, the building, and, on a busy day, up to 25,000 visitors. A sample of one day's activities appears on these pages.

For Kevin Johnson (1) the morning starts with ten minutes hard labour, as he winds the Wells Cathedral Clock, one of 70 which are kept running in the *Time Measurement* gallery. Six centuries ago, when this clock was new, the same task might have fallen to the cathedral's Sacrist.

Behind the scenes in *The Exploration of Space*, electronics technician Keith Gray (2) prepares automatic equipment which drives the exhibition's many audio-visuals. With a total of seven hundred working exhibits and audio-visuals to look after, the museum's maintenance team faces a formidable daily challenge.

In the Data-processing Centre, Indra Sookoo (3) checks print-out from a computer network which keeps a record of items in the collections and books in the library. Up-to-date information is available on-line to staff throughout the building.

1

2

3

School groups arrive throughout the morning, many of them to visit *Launch Pad*, an exhibition of experiments and demonstrations which anyone can try. Ann de Caires (4) is on duty there to greet them and to help younger visitors explore the exhibits in her charge.

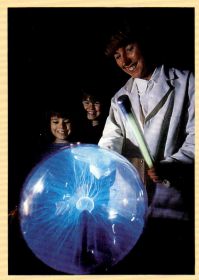

4

Engineering technician Mac McLauchlan (5) is preparing a sectioned aero engine for display. With his highly skilled colleagues in the museum workshop he tackles varied tasks which include conservation, model-making and the construction of working demonstrations rugged enough to withstand the enthusiastic use they will inevitably get in a popular museum.

For Joseph Cummings (6) the day's duty has been in the *Land Transport* galleries. Surrounded by historic fire engines, he gives advice to a visitor from Norway. Next week he will move to a new area, and with six floors and sixty galleries to cover, it could be twelve months before he returns to the fire engines. Security staff have round the clock responsibility for the building and its contents, and keep a watchful but friendly eye on three million visitors each year.

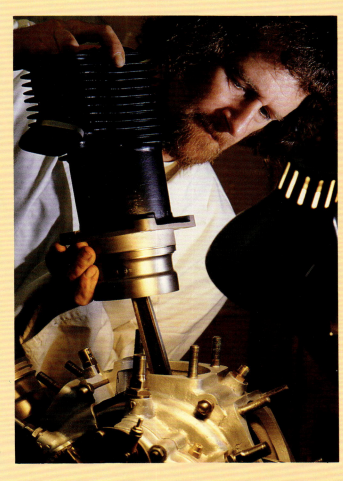

5

6

objects carry information, though it is often in a hidden form. Answers must be found to many of the questions which an object can raise: what is it? where did it come from? who made it, how and why? who used it and how did it work? what was its significance when it was made, and what significance does it have for us today? Research helps to give an object an identity, extracts from it information which can be of value to historians, and suggests meanings which can be attached to it when it is displayed.

Exhibited singly or in groups, alone or supported with explanatory information, objects can evoke a wealth of differing reactions. Visitors bring their own experience and

Small pieces of corroded brass (1), brought to the museum by a visitor in 1983, turned out to have surprising historical significance. Their arrival was the trigger for intensive detective work by curators who identified them as fragments from an instrument which was part sundial and part mechanical calendar. Research at home and abroad, supported by X-ray fluorescence analysis and examination by electron microscope, indicated that the sundial-calendar was 1500 years old. A reconstruction was made to show what the complete instrument may have looked like and how it worked (2). The original parts, now acquired by the museum, were put on display, and the research results published (3).

Because it made use of gearwheels, centuries before the first mechanical clocks, the sundial-calendar is historically important.

Its date places it at the centre of a thousand-year gap in the evidence. It suggests a way by which the ancient Greek tradition of sophisticated gear-wheel technology may have passed through the Byzantine empire to the Islamic world. From there it may have come to Western Europe and into our own clockmaking tradition. Modern clocks and watches may trace an ancestry to this 1500 year-old survivor.

2

3

expectations to the museum, and interpret the exhibitions in a host of individual ways.

The National Museum of Science and Industry serves five million visitors a year, more than half of them at the Science Museum in London. To each of these people it offers differing opportunities. Some come from curiosity, to find out how things happened in the past and how things are done today. Others come to feel the presence of the past, and so set their present lives in a new perspective.

For all who use it, the Science Museum can be a place where they reactivate an unfashionable emotion, the sense of wonder, as they share an extraordinary harvest of human ingenuity and achievement, preserved in collections which are without parallel in the world and openly displayed for all to see.

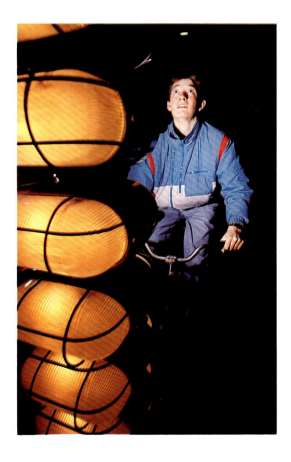

Converting muscle-power to electricity – an example of the new sort of museum experience available to participants in *Launch Pad*, the 'hands-on' exhibition at South Kensington. As its name implies, *Launch Pad* is a starting place for personal exploration of principles and phenomena, many of which have applications shown elsewhere in the museum.

Within the museum or out in the open, exhibits come alive when they can be shown in action. The sights and sounds of harvest time two generations ago are recreated for visitors to *Science Museum Wroughton* when a 1930s Marshall thresher goes through its paces (1). To show what rail travel was like in 'broad gauge' days – the early years of the Great Western Railway when the rails were seven feet apart – the Museum has constructed a full-scale reproduction of Daniel Gooch's famous *Iron Duke* of 1847 (2). Steamed at venues around the country, it provides an experience not otherwise available, for no operational broad gauge locomotives have survived.

List of illustrations

The following items from the museum's collections and from the holdings of the Science Museum Library are shown in this booklet. The figures in brackets are museum inventory numbers.